21

SUPER SIMPLE Physics
EXPERIMENTS

Rebecca W. Keller, PhD

REAL SCIENCE 4 Kids

Illustrations: Janet Moneymaker
                  Marjie Bassler

21 Super Simple Physics Experiments
ISBN 978-1-936114-93-1

Published by Gravitas Publications Inc.
www.gravitaspublications.com
www.realscience4kids.com

GRAVITAS
PUBLICATIONS

What Are Super Simple Science Experiments?

Super Simple Science Experiments are experiments that each focus on one aspect of scientific investigation. Doing science requires the development of different types of skills. These skills include the ability to make good observations, turning observations into questions and/or hypotheses, building and using models, analyzing data, using controls, and using different science tools including computers.

Super Simple Science Experiments break down the steps of scientific investigation so that you can focus on one aspect of scientific inquiry. The experiments are simple and easy to do, yet they are *real* science experiments that help you develop the skills needed for *real* scientific investigations.

Each experiment is one page long and lists an objective, the materials needed, a brief outline of the experiment, and any graphics or illustrations needed for the experiment. The skill being explored is shown in the upper right hand corner of each page.

The recommended companion book, *Super Simple Science Experiments Laboratory Notebook*, is a great place to record all the results of your experiments. It contains blank pages, lined pages, graph pages, and boxes for drawings.

Getting Started

On the next page is a list of the materials needed for all the physics experiments in this book. All the materials can be collected ahead of time and placed in a storage bin or drawer.

Materials at a Glance

Super Simple Science Experiments
    Laboratory Notebook
balloon
balls, two or more pairs of different types:
    basketball
    tennis ball
    ping pong ball
    baseball
bathtub
battery, AA
block, small
books, several, or 2 chairs
boulder or other large, heavy object
bowl
chair, 1-2
eggs, 4 hardboiled
magnets, 2 bar magnets with
    N and S poles marked
marble, 1 small and 1 large
marking pen
metal items such as:
    coins
    nail, iron
    nails, aluminum or steel
    paper clip
    other metallic items
objects, several heavy
paper
pencil, long
pencils, colored
plastic wrap
pole, steel
potatoes, 4.5 kg (10 lb.) bag
prism
rocks, 2
ruler
screw
spool, small
stopwatch
string, strong, about 2 m (6')
table
tape, strong adhesive
toy bucket
toy car

vegetable oil
wheelbarrow or wagon
wooden board or plank, .3 m x 1.2 m x
    2.5 cm thick (1' x 4' x 1")
wooden cove moulding, 1 m (3') long
wooden cove moulding, 3 pieces,
    each .3 m (1') long
watermelon, large

suggested items:
    tennis ball
    feather
    banana
    water balloon

available from Home Science Tools
(as of this writing)
www.hometrainingtools.com

    bulb, 1.5 volt, EL-LAMP1.5
    bulb holder, EL-BULBHD1
    alligator clip leads (2), EL-ALCLIP2
    battery holder for AA battery,
        EL-BATHAA1
    iron filings, CH-IRON

Table of Contents

Title   **Page**

1. What Goes Up Must Come Down

making observations

Objective

To observe whether or not different objects will fall to Earth when thrown.

Materials

suggested items:
    tennis ball
    marble
    crumpled piece of paper
    feather
    banana
    water balloon
Super Simple Science
    Experiments Laboratory Notebook

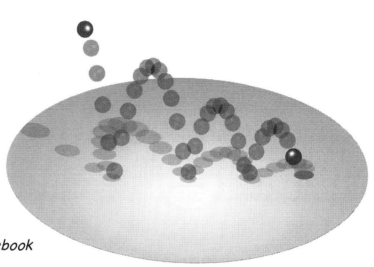

Experiment

❶ Write the name of each item in your *Laboratory Notebook*. Leave enough room below each item name to record your observations.

❷ Throw the item up in the air. Notice how it travels and then observe if it comes back down. Record your observations in your *Laboratory Notebook*.

❸ Answer the following questions in your notebook.

1. Do you think all objects will fall back to Earth when thrown?
2. Can you name any objects that might stay in the air when thrown?
3. Do you think you would observe the same result if you were on top of a tall mountain, near the ocean, or at the bottom of a valley?

Results and Conclusions

Gravity is a force that acts on all objects. When an object is thrown, it will move upwards for awhile until gravity takes over and pulls it back down. Even lightweight objects, such as a feather, will fall to the Earth.

2. Galileo's Experiment

Objective

To repeat Galileo's experiment and observe how objects of different weights fall to the ground.

Materials

two or more pairs of balls of different weight
    (basketball/tennis ball)
    (ping pong ball/tennis ball)
    (ping pong ball/baseball)
Super Simple Science Experiments
    Laboratory Notebook

Experiment

❶ Choose two balls of different types and write their names in your *Laboratory Notebook*.

❷ From a standing position, drop the two balls simultaneously from chest height. Observe whether one ball strikes the ground first or whether both balls strike the ground at the same time. Record your observations in your *Laboratory Notebook*.

❸ Repeat Steps ❶-❷ one or more times using two different balls each time. Record your results.

Results and Conclusions

Many early scientists believed that a heavier object will strike the ground before a lighter object. Galileo's experiment shows that objects of unequal weight will hit the ground simultaneously. This happens because Earth's *gravitational force* is the same for light objects as it is for heavy objects.

3. Inclined Planes

Objective

To observe how pushing an object up an inclined plane requires less force than lifting that same object straight up.

Materials

wood plank or board, .3 m x 1.2 m
x 2.5 cm thick (1′ x 4′ x 1″)
large watermelon
*Super Simple Science Experiments
    Laboratory Notebook*

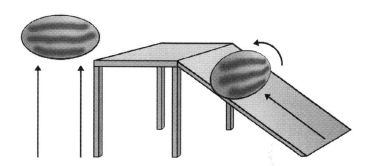

Experiment

❶ Keep your back straight as you do this experiment. Place the watermelon on the ground by bending your knees. Pick up the watermelon and try to lift it to the height of your waist by straightening your knees and keeping your back straight. Observe how easy or difficult it is to lift the watermelon. Record your observations in your *Laboratory Notebook*.

❷ Place one end of the wood plank on a table and the other end on the floor. Place the watermelon near the end of the plank that is on the floor and roll the watermelon up the plank to the top. Observe how easy or difficult it is to lift the watermelon using the plank. Record your observations.

Results and Conclusions

The *inclined plane* is one of six simple machines that are used to make doing work easier. Any flat surface that has one end lifted to any height can be used as an inclined plane. An inclined plane can be used to move heavy objects from one elevation to another. When using an inclined plane, the same amount of work is performed as when lifting an object; however, the force required to move an object from one elevation to another is distributed over a longer distance, making the object easier to move.

4. Acceleration

measuring data

Objective

To observe what happens when a marble travels down an inclined plane.

Materials

wooden cove moulding, 1 meter (3 feet) long
marble
stopwatch
marker
ruler
Super Simple Science Experiments
   Laboratory Notebook

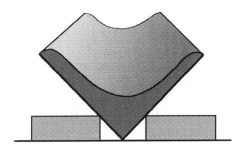

Experiment

❶ Calculate the distance of 1/4 of the length of the moulding. Starting at one end of the moulding, measure this distance and mark it in a way that your one-quarter point mark can easily be seen.

❷ Place one end of the moulding against a table or chair .3-.6 meter (1-2 feet) from the ground and the other end on the floor so that the one-quarter point is closest to the top. Prop up the two sides of the moulding to make a valley.

❸ Hold the marble at the top of the moulding and with the stopwatch ready, release the marble and start the stopwatch. When the marble reaches the end of the moulding, stop the stopwatch. Repeat several times and record the numbers in your *Laboratory Notebook*.

❹ Repeat the experiment, recording the moment the marble passes the one-quarter point. Compare your answers. Is the one-quarter point time half, longer, or shorter than the time required for the marble to travel the full distance?

Results and Conclusions

One of Galileo's most famous experiments uncovered how an object will accelerate down an inclined plane. Galileo discovered that a marble will travel one quarter of the distance down an inclined plane in half the time it takes to travel down the entire plane, meaning that the ball speeds up *(accelerates)* on the way down.

5. Rolling Down and Rolling Up

Objective

To observe what happens when a marble travels down and up two inclined planes.

Materials

3 pieces wooden cove moulding,
   each .3 m (1') long
marble
stack of several books – or 2 chairs
marker
ruler
strong adhesive tape
Super Simple Science Experiments
Laboratory Notebook

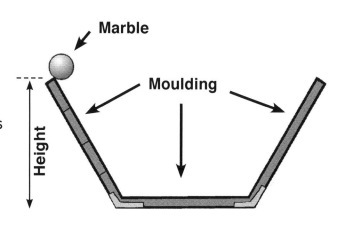

Experiment

❶ Assemble the wood mouldings as in the illustration above. Secure the undersides with adhesive tape, lining up the valleys. Rest each side on a stack of books or against a chair. On one of the elevated mouldings, mark the upper one-quarter point, the midpoint, and the lower one-quarter point.

❷ Hold the marble at the top of the moulding and measure the distance from the ground. Record this distance in your *Laboratory Notebook*. Release the marble and observe where on the opposite molding the marble stops. Mark this spot and measure the distance from the ground. Record this number.

❸ Repeat the experiment holding the marble at the upper one-quarter point, half-point, and lower one-quarter point. In each instance, record in your *Laboratory Notebook* the distance the marble travels up the opposite side.

Results and Conclusions

One of Galileo's well-known experiments revealed that an object that rolls down an inclined plane and up the side of an opposite inclined plane will reach its initial height. Does this illustrate *conservation of energy*? Why or why not?

6. Speedy Marbles

using equations

Objective

To calculate the speed of a marble.

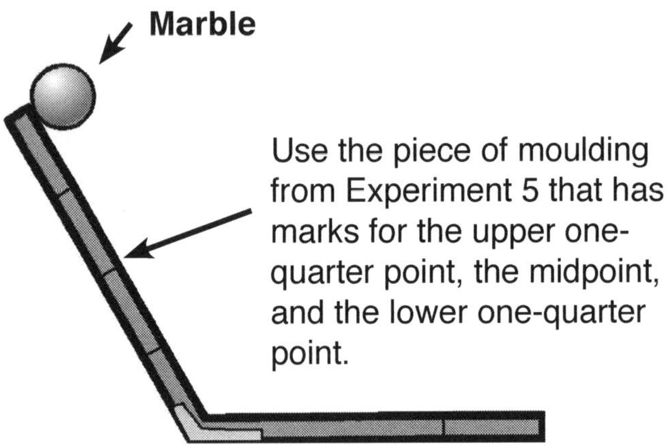

Marble

Use the piece of moulding from Experiment 5 that has marks for the upper one-quarter point, the midpoint, and the lower one-quarter point.

Materials

2 wooden cove mouldings.

marble

stack of several books or 1 chair

marker

ruler

stopwatch

strong adhesive tape

Super Simple Science Experiments Laboratory Notebook

Experiment

❶ Reassemble two of the wood trim mouldings from Experiment 5 to look like the illustration above. Secure the undersides with adhesive tape and rest the elevated side on a stack of books or against a chair. On the horizontal moulding measure and mark the one-quarter point from the unattached end.

❷ Release the marble at the top of the elevated moulding. When the marble hits the intersection point, start the stopwatch. When the marble passes the one-quarter point marked on the second moulding, stop the stopwatch and record the time in your *Laboratory Notebook*.

❸ Repeat the experiment three times, holding the marble at the upper one-quarter point, the midpoint, and the lower one-quarter point. In each instance, record the time the marble travels from the intersection to the far one-quarter point.

Results and Conclusions

You can calculate the speed of the marble by dividing the distance the marble travels over the time it takes it to travel. As the marble is released from lower and lower points, it has less energy, and as a result has less speed.

7. Marble in a Bowl

Objective

To observe the effect of centripetal force exerted on a marble.

Materials

1 small and 1 large marble
bowl
*Super Simple Science Experiments
Laboratory Notebook*

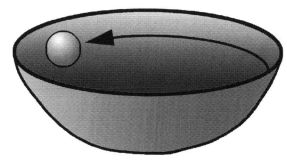

Experiment

❶ Place the small marble in the bowl.
What does it do? Record your observations in the *Laboratory Notebook*.

❷ Pick up the bowl, and start to gently move the bowl in a circular motion (as if you were swirling chocolate milk). What happens to the marble? Record your observations in the *Laboratory Notebook*.

❸ Repeat the experiment using the large marble. Notice any differences in how easy or difficult it is to get the large marble to move in the bowl. Does the large marble move faster or slower than the small marble? When you stop spinning the bowl, does the large marble stop sooner or later than the small marble?

Results and Conclusions

When you place a marble in a bowl and swirl the bowl, the marble will start to move in a circular path, following the sides of the bowl. The circular path of the bowl creates a *centripetal force* that acts on the marble. Centripetal means "center seeking." The centripetal force is actually pulling the marble towards the center of the bowl. The motion of the marble also creates an outward force, referred to as *centrifugal force*, that is contained by the sides of the bowl. When the inward and outward forces are balanced, the marble circles the inside of the bowl. (If the marble escapes from the top of the bowl, what happens?) The larger marble has more mass than the smaller marble and therefore has more *inertia* and more *momentum* once it is moving. It is harder to get the larger marble moving, but once it is moving, it takes longer for it to slow down.

8. Swinging Marble

making observations

Objective

To observe how potential energy is converted to kinetic energy.

Materials

marble
string, 0.5 meter (about 2 feet) long
adhesive tape
*Super Simple Science Experiments
    Laboratory Notebook*

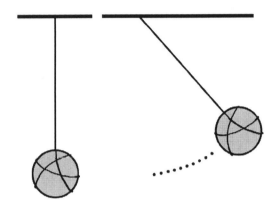

Experiment

❶ Wrap one end of the string around the marble and secure it with adhesive tape.

❷ Tape the free end of the string to the underside of a table or somewhere that when the marble is pulled, it can swing freely.

❸ Pull the marble up and outward and let it go. What happens? Note how far up you pulled the marble. Does the marble travel the same distance up on the other side? Try pulling the marble higher or lower. In each case, does the marble travel up to the same place on the other side? Record your answers and observations in your *Laboratory Notebook.*

Results and Conclusions

A marble attached to a string and hanging from a fixed point makes a pendulum. When the marble is pulled outward and upward and released, it will swing past the centerline to the other side. Just before the marble is released, it has *potential energy.* Potential energy is stored energy that has the potential to do work. Once the marble is released, the potential energy is converted to kinetic energy, the energy of motion.

9. Working Eggs

Objective

To observe how potential energy is used to do work.

Materials

4 hardboiled eggs
book
chair
table
Super Simple Science Experiments
   Laboratory Notebook

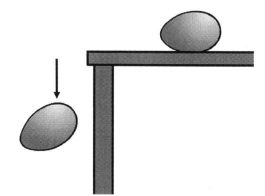

Experiment

❶ Place an egg on the floor. Without a chick inside to break the eggshell, do you think the egg can open on its own? Why or why not? How might you open the egg? Write your ideas in your *Laboratory Notebook*.

❷ Place an egg on a book that is sitting on the floor. Gently roll the egg off the book. Observe what happens. Record your observations.

❸ Place an egg on a chair. Gently roll the egg off the chair. Observe what happens. Record your observations.

❸ Place an egg on a table. Gently roll the egg off the table. Observe what happens. Record your observations, then go make egg salad.

Results and Conclusions

In the absence of a live chicken getting ready to hatch, an egg sitting on the floor will not likely open or move on its own. However, if you elevate the egg above the floor, the egg now has potential energy that can be converted to kinetic energy to do work. With enough potential energy, there may be enough kinetic energy that when the egg hits the floor, the shell breaks. The amount of work done on the egg is equal to the change in kinetic energy. Little change in kinetic energy (egg rolling off a book) may not break the egg. But large changes in kinetic energy (the egg falling from a high surface like the table) will cause the egg to break.

10. Lift Like Superman!

Objective

To observe how force is multiplied by using a lever and a fulcrum.

Materials

small block
sturdy wooden board, about 1.2 meters
    (4 feet) long
large, heavy object or boulder
Super Simple Science Experiments
    Laboratory Notebook

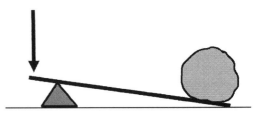

Experiment

❶ Place one end of the board *(lever)* under
a heavy object or boulder. Place a small
block *(fulcrum)* under the other end of the

lever so the fulcrum is close to that end of the lever.

❷ Push down on the end of the board nearest the fulcrum. Can you lift the large
object or boulder? Why or why not? In your *Laboratory Notebook,* record your
observations for this and the following steps of the experiment.

❸ Next, position the fulcrum near the middle of the board. Push on the end opposite
the large block or boulder. Can you lift the heavy object? Why or why not?

❸ Position the fulcrum close to the large block or boulder. Push on the end farthest
away from the fulcrum. Can you lift the heavy object? Why or why not?

Results and Conclusions

A *lever* with a *fulcrum* allows force to be multiplied. When a fulcrum is positioned
farthest from a heavy object *(the load)*, a large amount of *input force* (lots of
work to push the lever down) is needed to lift the heavy object. As the fulcrum is
moved closer to the load, the input force is multiplied. When the fulcrum is very
close to the load, a small amount of input force can be multiplied enough to lift a
heavy object.

11. Screws

Objective

To observe how a screw is really a rolled up inclined plane.

Materials

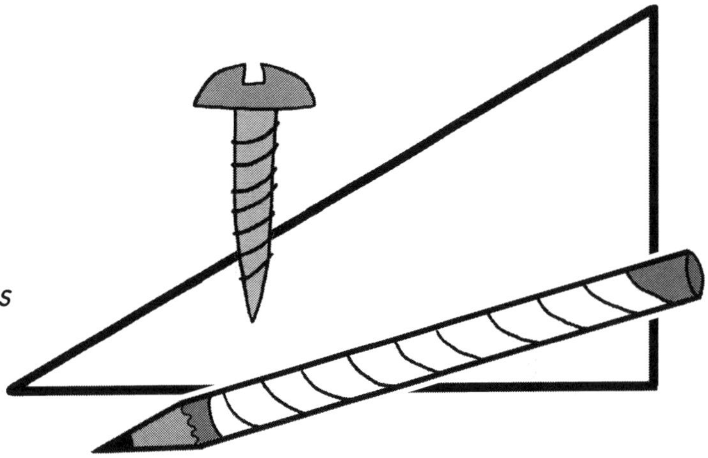

screw
paper
long pencil
tape (optional)
Super Simple Science Experiments
Laboratory Notebook

Experiment

❶ Take the screw and examine it carefully. Look at the sides and notice the ridges and how they are angled. Are they angled up or down? Write your observations in your *Laboratory Notebook*.

❷ Cut a paper triangle as shown in the drawing.

❸ Place the pencil on the edge of the paper. Wrap the paper triangle around the pencil by rolling the pencil. If you have trouble getting it started, try taping the paper to the pencil.

❹ Observe the pencil and paper. Do you see ridges along the sides? Are they similar to those on the screw? Why or why not?

Results and Conclusions

A screw is a simple machine that can be used to turn *rotational motion* into *linear movement*. Screws are commonly used as fasteners to hold objects together, but screws can be used to seal bottles and containers, move heavy objects uphill, and measure lengths with high accuracy.

12. Wheel and Axle

making observations

Objective

To observe how a wheel and axle can move a heavy load.

Materials

toy car
wheelbarrow or wagon
4.5 kg (10 lb.) bag of potatoes
Super Simple Science Experiments
    Laboratory Notebook

Experiment

❶ Observe how the toy car rolls. Flip it over and spin the wheels, then roll it a few times on the ground. Imagine how real cars work and how many people or objects they can carry. Record your observations and ideas in your *Laboratory Notebook.*

❷ Place the 4.5 kg (10 lb.) bag of potatoes on the ground. Keeping your back straight, lift the bag up and move it about 10 meters (10 yards). Imagine moving the potatoes 50 times back and forth. How much energy do you think it would take? Do you think you might get tired? Record your observations.

❸ Place the bag of potatoes in a wagon or wheelbarrow. Now move the potatoes about 10 meters (10 yards). Move the potatoes this distance several times. Do you think it would be easier to move the potatoes 50 times with a wagon or wheelbarrow than to carry them? Why or why not? Record your observations.

Results and Conclusions

A wheel and axle can be used to convert *rotational motion* to *linear movement.* The wheel is fixed to the axle, and when the wheel moves, it creates a mechanical advantage and multiplies force. This means that for a small applied force (which requires less energy), a large force is generated. In the case of a wheelbarrow or wagon, a small force applied to lift and push the wheelbarrow or pull the wagon generates enough force to move the 10 lb bag of potatoes with small energy input.

13. Pulley

Objective

To observe how a pulley can be used to move a heavy load.

Materials

small spool
long pencil
strong string, 1.2–1.5 meters (4–5 feet)
toy bucket
several heavy objects to go in the toy
   bucket
*Super Simple Science Experiments
   Laboratory Notebook*

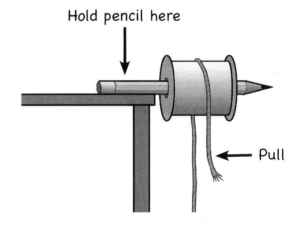

Hold pencil here

Pull

Experiment

❶ Attach one end of the string to the bucket. Place several heavy objects in the bucket and lift. Record your observations in your *Laboratory Notebook*.

❷ Assemble the spool, pencil, and string as shown in the drawing. Drape the free end of the string over the spool.

❸ Hold the pencil firmly against a tabletop and pull on the free end of the string. See if you can pull the bucket and objects upward. Is it easier to lift the objects in the bucket by using a pulley? Why or why not? Record your observations.

Results and Conclusions

A *pulley* is a type of simple machine that can be used to convert *rotational motion* to *linear movement*. A pulley is essentially a wheel and axle with a string across the wheel. In this example a simple, single pulley changes the direction of the force, splitting the force, making it easier to lift the load.

14. Friction

Objective

To observe how friction generates heat.

Materials

vegetable oil
Super Simple Science Experiments
   Laboratory Notebook

Experiment

❶ Place your palms together and quickly rub them back and forth. Observe whether or not your hands warm up. Record your observations in your *Laboratory Notebook*.

❷ Repeat Step ❶ and count how many times you rub your hands back and forth before they heat up. Record your observations.

❸ Now coat your palms with vegetable oil. Place your palms together and rub them again. Count how many times you rub them together before they heat up. Record your observations.

Results and Conclusions

On a chilly day, this simple experiment can warm your hands, arms, and legs. When two objects, like your hands, are moved back and forth briskly, they generate heat because of *friction*. When you apply oil to your hands, there is less friction between your palms and you should observe that it takes longer to generate heat.

15. Simple Circuits

Objective

To observe how an electric circuit works.

Materials

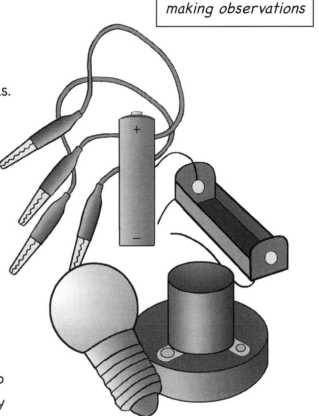

1.5 volt light bulb
light bulb holder
2 alligator clip leads
1.5 volt battery (AA)
AA battery holder
Super Simple Science Experiments
   Laboratory Notebook

Experiment

❶ Insert the light bulb into the light bulb
holder and the battery into the battery
holder.

❷ Take one of the alligator clip leads and
connect one end to one of the screw terminals of the light bulb holder. Connect
the other end to one wire of the battery holder. Take the other alligator clip lead
and fasten it to the other terminal of the light bulb holder and the other wire of
the battery holder. Record your observations in your *Laboratory Notebook*.

❸ Unhook one of the leads and observe what happens. Record your observations.

Results and Conclusions

Electricity requires a complete *circuit* to flow. When all of the leads are
connected, electricity from the battery can flow to and illuminate the light bulb.
When one of the leads is disconnected, electricity cannot flow from the battery to
the light bulb and the light bulb goes dark.

16. Sticky Balloons

making observations

Objective

To observe the attractive force of static electricity.

Materials

balloon
paper, hair, and a wall
Super Simple Science Experiments
  Laboratory Notebook

Experiment

❶ Inflate the balloon. Place the balloon on a wall or on a piece of paper and observe whether the balloon will stick to the wall or paper. Record your observations in your *Laboratory Notebook*.

❷ Rub the balloon in your hair until you feel your hair become attracted to the balloon.

❸ Place the balloon on a wall or a piece of paper and see if it will stick. Record your observations.

Results and Conclusions

In this experiment you explored how a balloon can pick up *electrons* from another object and how these electrons create an attractive force.

Electrons are negatively charged. When a balloon gains more electrons by picking them up from your hair, the balloon becomes negatively charged. When the negatively charged balloon is placed on a surface that is positively charged, the negatively charged electrons and the positively charged surface will attract and the balloon will stick. The balloon may also stick to a surface that has a lesser negative charge (a surface that has fewer free negatively charged electrons than the balloon has).

17. Magnetic Fields

Objective

To observe magnetic fields.

Materials

sheet of white paper
bar magnet
iron filings
Super Simple Science Experiments
    Laboratory Notebook

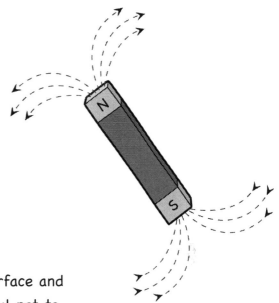

Experiment

❶ Place the sheet of white paper on a flat surface and gently sprinkle iron filings on it, being careful not to breathe them in.

❷ Slide the bar magnet underneath the sheet of paper and observe how the iron filings move. Record your observations in your *Laboratory Notebook*.

❸ Allow the bar magnet to sit for a few minutes and note how the iron filings align. Record your observations, including a drawing of the pattern the iron filings make.

Results and Conclusions

Magnets create a field of force called a *magnetic field* that can be easily observed using a bar magnet and iron filings, which are attracted to a magnetic field. Magnets have oppositely charged poles that are labeled north (N) and south (S). Poles that are alike repel each other, and opposite poles attract each other. Electric charges can exist by themselves, but magnetic poles can only exist as a pair. If you break the bar magnet in tiny pieces, each piece will still have a north pole and a south pole and each piece will generate a magnetic field.

18. Fun With Magnetic Fields

making observations

Objective

To observe how magnetic fields interact.

Materials

sheet of white paper
2 bar magnets with north and south poles marked
iron filings
Super Simple Science Experiments Laboratory Notebook

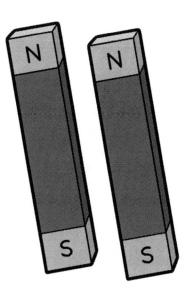

Experiment

❶ Place the sheet of white paper on a flat surface and gently sprinkle iron filings on it, being careful not to breathe them in.

❷ Slide one bar magnet underneath the sheet of paper and observe how the iron filings move. Record your observations in your *Laboratory Notebook*.

❸ Slide the second bar magnet underneath the white paper so that the north poles are side by side. Bring the magnets close to each other. Observe how the iron filings move. Record your observations.

❹ Flip one of the magnets so the north pole of one magnet is next to the south pole of the other magnet. Bring the magnets close to each other. Observe how the iron filings move. Record your observations.

Results and Conclusions

Magnetic fields can interact with other magnetic fields. The interaction of the magnetic fields will differ depending on whether the poles of each magnet are opposite each other or nearest each other.

19. Making Magnets

Objective

To observe which objects will become magnetic.

Materials

bar magnet
plastic wrap
iron filings
metal items such as:
  iron nail
  coins
  aluminum or steel nails
  metal paper clip
  other metallic items of your choice
Super Simple Science Experiments
  Laboratory Notebook

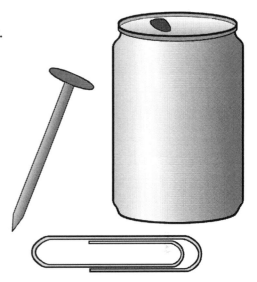

Experiment

❶ Wrap the bar magnet in plastic wrap.

❷ Put some iron filings on a piece of paper, being careful not to breathe them in. Bring the wrapped magnet close to the iron filings. Does the magnet pick up the filings? Record your observations in the *Laboratory Notebook*.

❸ Take the bar magnet out of the plastic wrap and place it next to the iron nail for a few minutes. Wrap the iron nail in plastic wrap and bring it close to the iron filings. Does the iron nail pick up the filings? Record your observations.

❹ Repeat Step ❸ with other metal items. Notice which items become magnetized and which don't. Record your observations.

Results and Conclusions

Magnetic fields will pass through most metals. However, iron is a metal that can become magnetized by being close to a magnet. The atoms in an iron nail will align to the magnetic field, and over time the nail will become a magnet itself.

20. Splitting Light

making observations

Objective

To observe what happens when a prism is used to split white light.

Materials

prism
sunny day
colored pencils
Super Simple Science Experiments
   Laboratory Notebook

Experiment

❶ On a sunny day take the prism outside. Hold the prism in your hands so that sunlight shines on it. Rotate it until you see it create colors on the ground below. Record your observations in your *Laboratory Notebook*.

❷ Continue to rotate the prism and observe how the colored light begins to reflect. Does it go up? Does it go down? How many colors do you see? Record your observations.

Results and Conclusions

White light, or sunlight, is actually a mixture of different colors. When sunlight is allowed to pass through a prism, the glass or plastic in the prism bends the light and separates the white light into the individual wavelengths it is made up of, creating a rainbow of many colors.

21. Exploring Sound

Objective

To play with sound and observe what happens as sound travels through different materials.

Materials

two rocks
a bathtub
a steel pole
Super Simple Science Experiments
   Laboratory Notebook

Experiment

❶ Have a friend stand a couple of meters (a few feet) away and hold the two rocks. Have them smash the rocks together until you hear the sound the rocks make. How loud was the sound? Record your observations in your *Laboratory Notebook*.

❷ Put your ear against a steel pole and tap one of the rocks against the pole and observe the sound. Was it louder or softer than when the rocks were banged together? Record your observations.

❷ Climb into a bathtub full of water and submerse your ears. Tap one of the rocks on the side of the tub and observe the sound. Was it louder or softer than in air or through the pole? Record your observations.

Results and Conclusions

Sound travels in *waves,* and therefore it will travel differently through different materials. Some materials carry sound better than other materials, and sound will be louder when traveling through these materials. The speed sound travels is also affected by the material it is passing through. Sound will travel about four times faster in water than in air and about 15 times faster in steel than in air.

More REAL SCIENCE-4-KIDS Books
by Rebecca W. Keller, PhD

Focus Series unit study program — each title has a Student Textbook with accompanying Laboratory Workbook, Teacher's Manual, Study Folder, Quizzes, and Recorded Lectures

Focus On Elementary Chemistry
Focus On Elementary Biology
Focus On Elementary Physics
Focus On Elementary Geology
Focus On Elementary Astronomy

Focus On Middle School Chemistry
Focus On Middle School Biology
Focus On Middle School Physics
Focus On Middle School Geology
Focus On Middle School Astronomy

Focus On High School Chemistry

Building Blocks Series yearlong study program — each Student Textbook has accompanying Laboratory Notebook, Teacher's Manual, Lesson Plan, and Quizzes

Exploring the Building Blocks of Science Book K (Activity Book)
Exploring the Building Blocks of Science Book 1
Exploring the Building Blocks of Science Book 2
Exploring the Building Blocks of Science Book 3
Exploring the Building Blocks of Science Book 4
Exploring the Building Blocks of Science Book 5
Exploring the Building Blocks of Science Book 6
Exploring the Building Blocks of Science Book 7
Exploring the Building Blocks of Science Book 8

Super Simple Science Experiments Series

21 Super Simple Chemistry Experiments
21 Super Simple Biology Experiments
21 Super Simple Physics Experiments
21 Super Simple Geology Experiments
21 Super Simple Astronomy Experiments
101 Super Simple Science Experiments

Kogs-4-Kids Series interdisciplinary workbooks that connect science to other areas of study

Physics Connects to Language
Biology Connects to Language
Chemistry Connects to Language
Geology Connects to Language
Astronomy Connects to Language

Note: A few titles may still be in production.

Gravitas Publications Inc.
www.realscience4kids.com

Made in the USA
Columbia, SC
16 March 2018